Interactive
Mathematics Program®

INTEGRATED HIGH SCHOOL MATHEMATICS

How Much? How Fast?

FIRST EDITION AUTHORS:
Dan Fendel, Diane Resek, Lynne Alper, and Sherry Fraser

CONTRIBUTORS TO THE SECOND EDITION:
Sherry Fraser, IMP for the 21st Century
Jean Klanica, IMP for the 21st Century
Brian Lawler, California State University San Marcos
Eric Robinson, Ithaca College, NY
Lew Romagnano, Metropolitan State College of Denver, CO
Rick Marks, Sonoma State University, CA
Dan Brutlag, Meaningful Mathematics
Alan Olds, Colorado Writing Project
Mike Bryant, Santa Maria High School, CA
Jeri P. Philbrick, Oxnard High School, CA
Lori Green, Lincoln High School, CA
Matt Bremer, Berkeley High School, CA
Margaret DeArmond, Kern High School District, CA

Key Curriculum Press

Second Edition   I M P

I0030829

This material is based upon work supported by the National Science Foundation under award numbers ESI-9255262, ESI-0137805, and ESI-0627821. Any opinions, findings, and conclusions or recommendations expressed in this publication are those of the authors and do not necessarily reflect the views of the National Science Foundation.

Key Curriculum Press
1150 65th Street
Emeryville, California 94608
email: editorial@keypress.com
www.keypress.com
10 9 8 7 6 5 4 3 2 1   15 14 13 12 11
ISBN 978-1-60440-147-9
Printed in the United States
of America

Project Editors
Mali Apple, Josephine Noah

Project Administrator
Emily Reed

Professional Reviewers
Rick Marks, Sonoma State University, CA
D. Michael Bryant, Santa Maria High School, CA, retired

Accuracy Checker
Carrie Gongaware

First Edition Teacher Reviewers
Kathy Anderson, Aptos High School, CA
Dan H. Brutlag, Tamalpais High School, CA
Robert E. Callis, Hueneme High School, CA
Susan Schreibman Ford, Delhi High School, CA
Mary L. Hogan, Arlington High School, MA
Jane M. Kostik, Patrick Henry High School, MN
Brian Lawler, California State University San Marcos, CA
Brent McClain, Vernonia School District, OR
Michelle Novotny, Eaglecrest High School, CO
Barbara Schallau, East Side Union High School District, CA
James Short, Oxnard Union High School District, CA
Kathleen H. Spivack, Wilbur Cross High School, CT
Linda Steiner, Orange Glen High School, CA
Marsha Vihon, Corliss High School, IL
Edward F. Wolff, Arcadia University, PA

First Edition Multicultural Reviewers
Genevieve Lau, Ph.D., Skyline College, CA
Luís Ortiz-Franco, Ph.D., Chapman University, CA
Marilyn Strutchens, Ph.D., Auburn University, AL

Copyeditor
Brandy Vickers

Interior Designer
Marilyn Perry

Production Editor
Andrew Jones

Production Director
Christine Osborne

Editorial Production Supervisor
Kristin Ferraioli

Compositors
Kristin Ferraioli, Maya Melenchuk

Art Editor/Photo Researcher
Maya Melenchuk

Technical Artists
Lineworks, Inc., Maya Melenchuk,
Kristin Ferraioli

Illustrator
Juan Alvarez, Alan Dubinsky, Tom Fowler,
Nikki Middendorf, Briana Miller, Evangelia
Philippidis, Paul Rodgers, Sara Swan, Martha
Weston, April Goodman Willy, Amy Young

Cover Designer
Jenny Herce

Printer
Lightning Source, Inc.

Executive Editor
Josephine Noah

Publisher
Steven Rasmussen

CONTENTS

How Much? How Fast?—Accumulated Change, Rates of Growth, and the Fundamental Theorem of Calculus

How Much? How Fast?

Accumulated Change, Rates of Growth,
and the Fundamental Theorem of Calculus

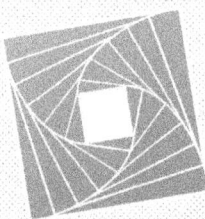

How Much? How Fast?—Accumulated Change, Rates of Growth, and the Fundamental Theorem of Calculus

Adding Up the Parts

When a quantity is changing, we often want to know two things: *At any given point, how fast is the quantity changing?* and *How much of it is there?* These two questions turn out to be closely related. Understanding this relationship is the main goal of this unit.

You'll investigate these and other ideas in the context of finding the volume of a pyramid, studying speed and distance, and solving a problem about a solar energy collector. You'll begin by focusing on the second question, *How much of it is there?*

Ian Carr and Halle Pozos work on a strategy for finding the volume of a pyramid.

Building the Pyramid

The ancient tombs of Egypt were built in the shape of a pyramid. Generally, the base of the pyramid was a square, and the sloping sides were triangles that met in a point above the center of the square.

Suppose such a pyramid was 100 feet high, with a square base that was 100 feet on each side. One of the questions you will investigate over the course of this unit is this:

What is the total volume of the pyramid?

For now, imagine that you want to build a pyramid like the one described above, but the only building blocks you have are cubes that are 1 foot on each side. You won't be able to create sloping sides like a genuine pyramid, so your goal is simply to build a good approximation.

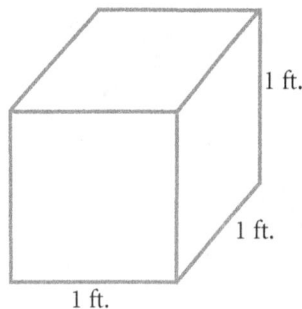

A building block

How will you build your approximation, and how many blocks will you need?

How Far Did You Go?

Suppose your new car has a device that makes a graph of your speed. One day, you start at noon and drive for 5 hours, varying your speed systematically. The graph of the car's speed is shown here.

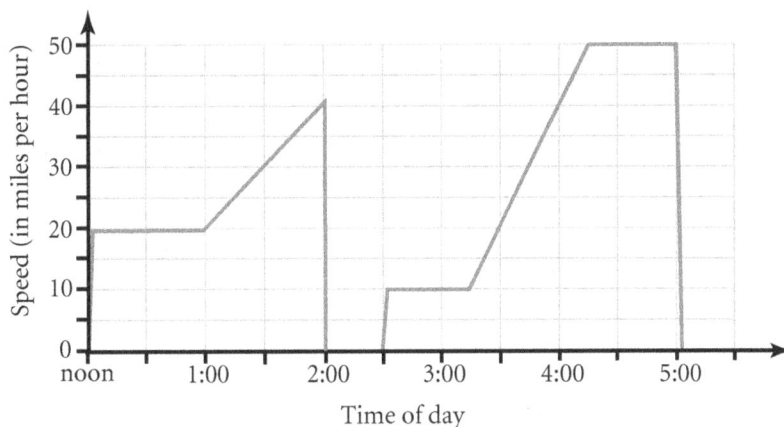

How many miles did you travel on your trip? Describe how you determined the distance.

Another Trip

This graph shows your new car's speed, as a function of time, for another trip.

How far did you travel altogether? Explain your reasoning.

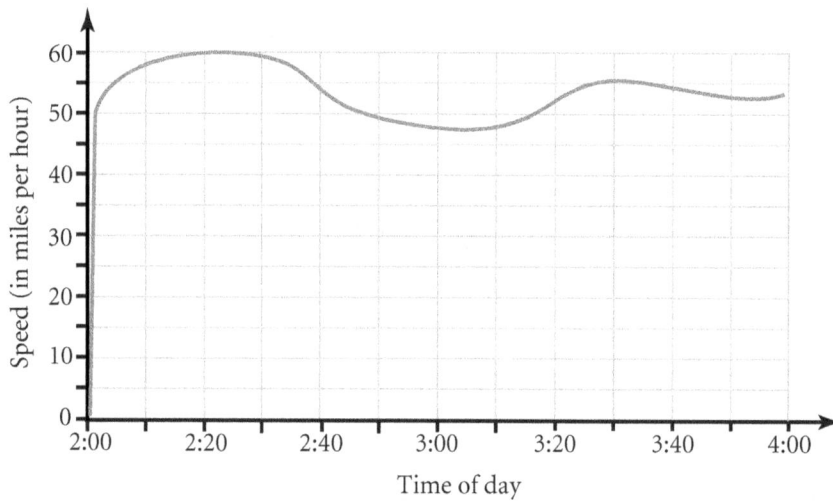

Advanced Pool Pockets

Imagine a modified pool table in which the only pockets are those in the four corners.

The diagram shows such a table as viewed from above.

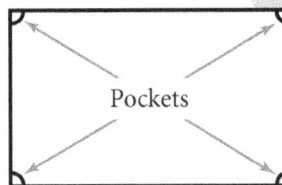

This POW will use the view from above, with different parts of the table labeled as shown.

Imagine that a ball is hit from the lower-left corner in a diagonal direction that forms a 45° angle with the sides.

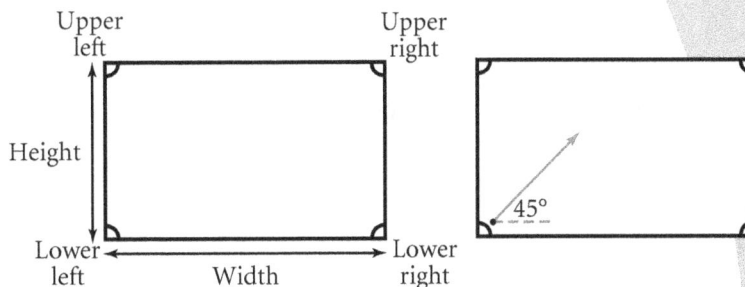

Every time the ball hits an edge of the table, it bounces off, again at a 45° angle, and continues this way until it hits one of the corner pockets perfectly.

For example, using the previous diagram, the first few bounces of the ball would look as shown here:

Of course, the ball's path will depend on the shape of the table. For example, with a square table, the ball would go directly into the opposite corner without bouncing at all.

Your task is to investigate what happens to the ball and how this depends on the dimensions of the table.

You should make these assumptions:

- The ball is always shot at an angle of 45°.
- The ball is always shot from the lower-left corner.
- The ball will keep traveling indefinitely, stopping only when it hits a corner.
- The ball behaves as an ideal point. That means it always bounces at an exact 45° angle, and it must hit a corner point exactly in order to stop.

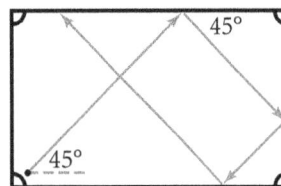

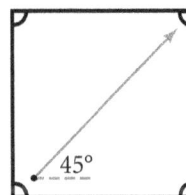

continued

In your investigation, you might look at these questions:

- Does the ball always hit a pocket eventually?
- If so, which pocket does it hit?
- If it does hit a pocket, how many times does it bounce before it hits the pocket?
- Do you get different kinds of results for different cases?
- What if the dimensions of the pool table are not whole numbers?

You may have some other questions of your own you'd like to investigate.

You will probably find it useful to have grid paper to record your investigation and to use the width of the squares on the grid paper as your unit of length. That way, you'll be able to keep track more easily of the ball's exact path.

For example, the diagram illustrates how you might use grid paper to show the ball's path for a 3-by-5 table. The dashed lines represent the lines of the grid paper, and the solid rectangle is the outline of the table. The diagonal lines with arrows show the path of the ball. At the end of the path, the ball is about to go into the pocket in the upper-right corner.

Happy bouncing!

Write-up

1. *Problem Statement*

2. *Process*

3. *Solution:* Considering various questions, state the most general solution you can find for each case you consider and for the problem as a whole. Justify your solutions as fully as you can.

4. *Self-assessment*

From *Mathematics: A Human Endeavor,* 2nd ed., by Harold R. Jacobs. (New York: W. H. Freeman and Company, 1970). Used with permission.

How Fast? How Much?

In this activity, you will explore the relationships between rates and amounts.

1. The situations presented in parts a to g each involve a constant rate. For each part, do these things:
   - Specify the rate and its units.
   - Name the variable quantities and their units.
   - Write an equation expressing one variable as a function of the other.
   - Use your equation to solve the problem.

   a. Lindsay automatically deposits 15% of her wages each month in a savings account. She earns $1760 per month. How much will she save in a year and a half?

   b. The winner of a turtle race covered the 5-meter course in 7 minutes 13 seconds. If the winner continues at this rate, how long would it take this turtle to travel 8 meters?

   c. A weight-loss center boasts that their diet will help people lose 25 pounds in 10 weeks. If they're right, and if the rate of weight loss is constant, how long would it take someone on the diet to lose 18 pounds?

   d. A movie theater charges $6.50 for a child's ticket. How much would it cost for five children to see a movie?

   e. A 1.5-liter bottle of a certain type of juice contains 210 grams of sugar. If you drink 0.4 liter of this juice (about 12 ounces), how much sugar would you consume?

   f. A high school student works three 6-hour shifts at a restaurant each week. How many weeks would the student have to work to accumulate 100 hours?

   g. A gallon of a certain type of paint covers approximately 450 square feet of surface area. Robin wants to paint the walls and ceiling of a rectangular room measuring 8 feet by 10 feet with a 7-foot ceiling. How much paint will he need?

continued ▶

2. Make up two more problems like this, using contexts of your own. One of your problems should involve some quantity changing over time. The other should use a different independent variable (not time).

3. All of the equations you wrote fit the same general pattern, or form. Describe this common form.

4. The graphs of your equations would also all have the same form. Choose any two of the equations and graph them, carefully labeling the quantities and their units. Describe this form, and explain how the problems could be solved directly from the graphs.

Leaky Faucet

The kitchen faucet has started dripping. For the first 40 minutes, it drips at a rate of 10 milliliters per minute. Then it gets worse, and for the next 30 minutes, it drips at a rate of 20 milliliters per minute. And then it gets much worse: for another hour and 20 minutes, it drips at 50 milliliters per minute. A plumber finally arrives and fixes the leak.

You noticed the problem right away and decided to set a 5-gallon bucket under the faucet to catch the dripping water.

1. Create a graph showing the rate at which the water is dripping.

2. Create a graph showing the amount of water in the bucket as a function of time.

3. Considering the data about the rate of the drip, describe how your two graphs are related.

4. Considering the amount of water in the bucket at any given time, describe how your two graphs are related.

Units for Measuring Electricity

The measurement of electricity can be confusing. We can't visualize either the thing being measured or the unit of measurement—unlike, say, the volume of water in a pond or the length of a fishing pole.

Electricity is a form of energy, and a certain amount of electricity can be thought of as a "packet" of energy. The basic scientific unit for measuring the *amount* of energy is the *joule* (J).

With electricity, it's sometimes easier to think about the rate at which a device consumes or produces energy. For instance, an electric clock might use 18,000 *joules per hour* (J/hr) of energy, while an LED nightlight might use 1000 joules per hour. Another term for rate of energy use is *power*.

To complicate things, electrical energy use is commonly measured with yet another unit. The *power* of an electrical device—that is, the *rate* at which the device consumes or produces energy—is measured in *watts*. One watt is equivalent to 1 joule per second, or 3600 joules per hour. In other words, a watt describes an amount of energy per unit of time.

For example, the electric clock mentioned previously consumes 5 watts of power. Some ordinary lightbulbs are 100-watt bulbs, which means they use 100 joules of energy per second. The rate of power usage for a household is typically measured in *kilowatts* (thousands of watts), while a power plant might produce energy at a rate measured in *megawatts* (millions of watts).

The amount of electricity used by an object like a lightbulb depends on both its rate of energy usage (that is, its power) and the length of time it operates. For instance, if a 100-watt bulb is on for 3 hours, it consumes 300 *watt-hours* of energy. If a 200-watt bulb is on for 10 hours, it consumes 2000 watt-hours, or 2 kilowatt-hours. These values could be converted to joules, but watt-hours is the more common unit.

What's Watt?

In a certain household, one 60-watt lamp is left on throughout the day and night. At 7 a.m., three 100-watt lamps are turned on, and they are turned off at 3 p.m. At 9 a.m., five 150-watt lamps are turned on, and they stay on until 6 p.m. At 8 p.m., two 200-watt lamps are turned on, and they are turned off at 10 p.m.

1. Make a graph showing the rate at which energy is being consumed by these lamps altogether over the course of a day. On the horizontal axis, use a scale that starts at midnight of one day and continues until midnight of the next day.

2. Find the total amount of energy consumed by the lamps over a 24-hour period.

Electrical Meter

The activity *What's Watt?* describes the electricity usage from a set of lamps in a particular household. As the day goes by, an electrical meter for the house registers the total amount of electricity these lamps have used so far.

Here is the information about the lamps:

- One 60-watt lamp is left on throughout the day and night.
- At 7 a.m., three 100-watt lamps are turned on. They are turned off at 3 p.m.
- At 9 a.m., five 150-watt lamps are turned on. They stay on until 6 p.m.
- At 8 p.m., two 200-watt lamps are turned on. They are turned off at 10 p.m.

Make a graph showing the meter reading as a function of time, beginning at midnight and continuing for 24 hours. Assume the meter was set to 0 at midnight, at the start of the 24-hour period.

Activity

Tilted Duct

A device for measuring rainfall consists of a vertical intake duct attached to a holding tank. The cross section of the duct is a square that is 5 centimeters on each side.

Two of these collectors are located side by side in a meadow. One is intact, but the intake duct of the other has been bent sideways. Both holding tanks were emptied yesterday. Today, after a steady, calm rain (no wind), the intact collector is found to contain 30 cubic centimeters (30 cc) of water.

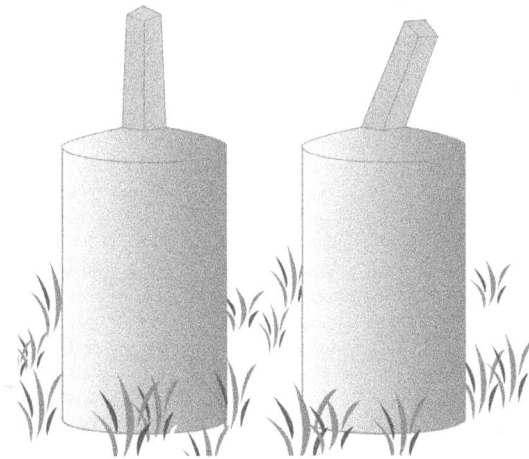

1. Suppose the bent duct is tilted at an angle of 10° from vertical. How much water would you expect to be in that collector?

2. Suppose the bent duct is tilted at an angle of 25° from vertical. How much water would you expect to be in that collector?

3. Make up two more angles, and determine how much water would have been collected in each case.

4. Make a conjecture about an equation to calculate the amount of water collected as a function of the angle of tilt.

Warming Up

A solar-collection system takes in energy during daylight hours. The rate at which the system accumulates energy depends on the way the sun's rays hit the solar panels.

At certain places on the earth during the spring and fall equinoxes, the sun rises at exactly 6 a.m. and sets at exactly 6 p.m. The maximum rate of energy accumulation occurs when the sun is most directly overhead (at noon) and the solar panels are oriented so that they receive these rays directly (that is, perpendicularly).

During the rest of the daylight hours, the panels get somewhat indirect sunlight, depending on the angle of the sun in the sky, as illustrated here. From 6 p.m. of one day until 6 a.m. of the next day, the system does not take in any energy, because there is no sunlight.

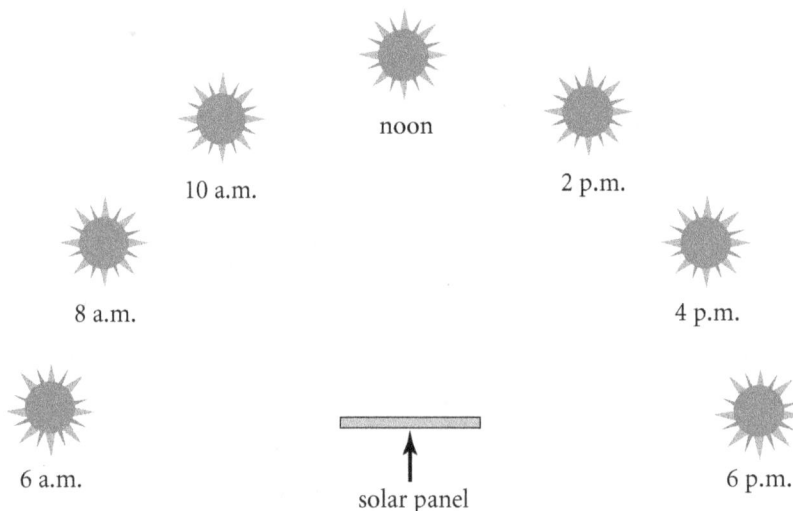

Assume the day is clear, so the sun is shining throughout the daylight hours. Also assume that at noon, the sun's intensity is such that a 1-foot-square solar panel is absorbing energy at a rate of 80 watts. The solar-collection system contains 30 square feet of panels and is therefore absorbing 2400 watts of energy.

Draw a graph showing the rate at which the system will absorb energy, over the course of a day, from midnight to midnight.

Total Heat

1. Based on the graph you created for *Warming Up*, estimate the total amount of energy absorbed by the solar-collection system during each of these periods.

   a. Midnight to 6 a.m.

   b. 6 a.m. to 8 a.m.

   c. 8 a.m. to 10 a.m.

   d. 10 a.m. to noon

   e. Noon to 2 p.m.

   f. 2 p.m. to 4 p.m.

   g. 4 p.m. to 6 p.m.

   h. 6 p.m. to midnight

2. Make a table for your results from Question 1. Include a column showing, for each time period, the total amount of energy absorbed from midnight to the end of that time period.

3. What is your estimate of the total amount of energy collected during one 24-hour period?

Rate and Accumulation

You're beginning to see some connections between the two big ideas in calculus: rate of change (*How Fast?*) and amount of accumulation (*How Much?*). You'll now learn to use derivatives, which you first studied in *Small World, Isn't It?*, to tie these two ideas together.

Sierena Parker and Shafari Jackson collaborate on making a rate graph from an accumulation graph.

How Fast Were You Going?

Your new car (from *How Far Did You Go?*) can also make an **accumulation graph** of the distance you have traveled as a function of time.

On one trip, you start the graph from the moment you turn onto the main road. The graph looks like this after an hour of driving. For example, as of 12:20 p.m., you have traveled a total of 10 miles.

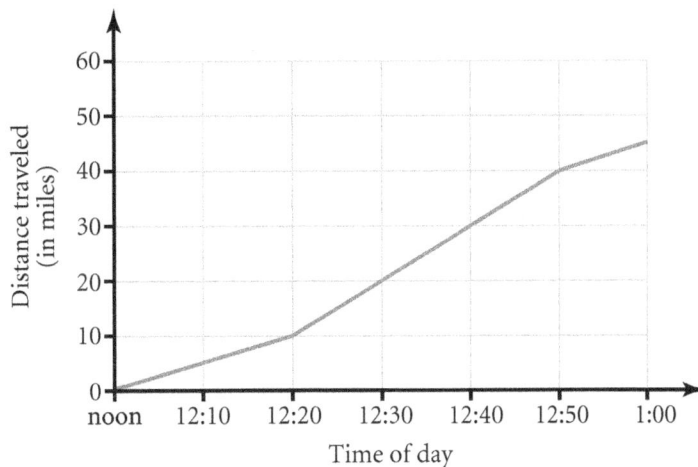

1. Find your car's average speed for the trip.

2. Using the same scales as this graph, make a **rate graph** showing the speed you were traveling, as a function of time, from noon until 1:00 p.m.

3. The graph shown above consists of several line segments. What does this tell you about your speed?

A Distance Graph

Here is a distance graph for a trip much shorter than the one in *How Fast Were You Going?*

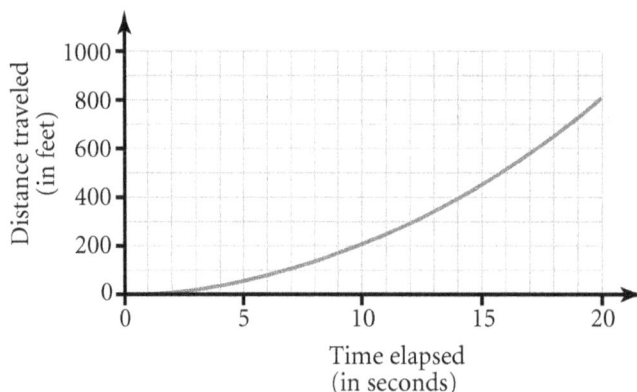

1. Find your car's average speed over this 20-second time period.

2. Let t represent the time elapsed (in seconds). Use the graph to estimate the speed at which you were traveling, in feet per second, at each of these precise moments.

   a. $t = 5$

   b. $t = 10$

   c. $t = 15$

3. Estimate the speed for at least three more moments during the trip. Then plot those results and your results from Question 2 to show speed as a function of time.

Let It Fall!

A helicopter is ready to drop a supply bundle to a group of firefighters who are behind the fire lines.

The principles of physics that describe the behavior of falling objects state that when an object is falling freely, it goes faster and faster as it falls. In *High Dive,* you worked with a specific formula to describe this motion:

If an object is dropped from rest, then d(t), the distance it falls during the first t seconds after being dropped, is given by the equation d(t) = 16t².

1. Assume the bundle is dropped from a height of 400 feet, so it hits the ground in 5 seconds.

   a. Justify the statement that the bundle hits the ground in 5 seconds.

   b. What is the bundle's average speed during the 2 seconds before it hits the ground (that is, during the 2-second interval from $t = 3$ to $t = 5$)?

   c. What is the bundle's average speed during the last second before it hits the ground?

   d. What is the bundle's average speed during the last tenth of a second before it hits the ground?

   e. What is the bundle's average speed during the last hundredth of a second before it hits the ground?

   f. What do you think is the bundle's speed at the exact moment it hits the ground?

continued ▶

2. The answer to Question 1f is the bundle's **instantaneous speed** at $t = 5$. Now consider $t = 2$.

   a. Find the bundle's average speed during the half second before $t = 2$, that is, from $t = 1.5$ to $t = 2$.

   b. Find the bundle's average speed during the tenth of a second before $t = 2$, that is, from $t = 1.9$ to $t = 2$.

   c. Find the bundle's average speed during the hundredth of a second after $t = 2$, that is, from $t = 2$ to $t = 2.01$.

   d. What is the bundle's instantaneous speed at $t = 2$?

3. Find the bundle's instantaneous speed for some other value of t less than 5.

Basic Derivatives

For simple functions, you can use algebra to find the **derivative** function. To do so, you will use the definition that for a function $y = f(x)$, the derivative of f at $x = a$ is the value approached by the ratio

$$\frac{f(a + h) - f(a)}{h}$$

as h gets smaller and smaller.

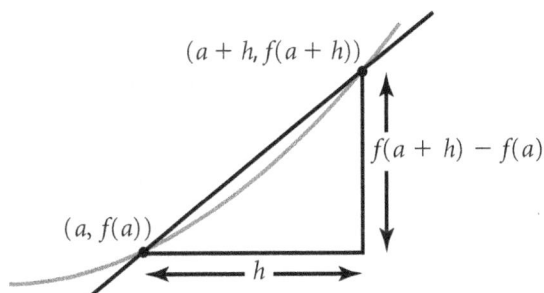

This value is written as $f'(a)$, which is read "f prime of a."

1. Consider the function $f(x) = x^2$.

   a. Use the definition of the derivative to find $f'(1), f'(2), f'(3)$, and $f'(4)$.

   b. Look for a pattern to develop a general rule for the value of $f'(a)$.

   c. Justify your general rule using algebra.

2. Consider the function $g(x) = x^3$.

   a. Use the definition of the derivative to find $g'(1), g'(2), g'(3)$, and $g'(4)$.

   b. Look for a pattern to develop a general rule for the value of $g'(a)$.

   c. Justify your general rule using algebra.

continued ▶

3. Consider the function $h(x) = x^4$.

   a. Use the definition of the derivative to find $h'(1), h'(2), h'(3)$, and $h'(4)$.

   b. Look for a pattern to develop a general rule for the value of $h'(a)$.

   c. Justify your general rule using algebra.

4. Consider the function $k(x) = 5x$.

   a. Use the definition of the derivative to find $k'(1), k'(2), k'(3)$, and $k'(4)$.

   b. Explain why you could have expected the results you found in Question 4a.

5. Consider the function $p(x) = 5x^2$.

   a. What do you conjecture for the derivative $p'(x)$?

   b. Test your conjecture using algebra.

Summer Job

Jessie and Rex find jobs at a summer camp. They start on the same day, working the same hours and earning the same wages, $9.50 an hour. Both are saving for college, so they arrange for their employer to deposit 60% of their earnings directly into their savings accounts, which they vow not to deplete.

On the day they begin work, Jessie's savings account contains $280. Rex's contains $350.

1. a. Write an equation showing Jessie's savings account balance, $J(h)$, as a function of the number of hours, h, she works.

   b. Do the same for Rex's savings account balance, $R(h)$.

2. Graph both equations on a single set of axes.

3. Your equations for Question 1 are really accumulation functions, so their derivatives should be the corresponding rate functions.

   a. Write an equation for the derivative $J'(h)$. Explain what it means in the context of the problem.

   b. Do the same for $R'(h)$.

4. Graph the two derivative functions on a single set of axes.

5. Describe and explain the various relationships you observe among your four graphs.

Going Up?

1. Gabriel likes to ride the escalator at the mall. Except he doesn't simply ride—he walks quickly up the steps of the escalator as the escalator rises.

   At $t = 0$, Gabriel gets onto a certain step S of the escalator as it starts going up. The graph of $h = f(t)$ shows how high step S has risen after t seconds. The graph of $h = g(t)$ shows how high Gabriel would have risen after t seconds if the escalator had not been moving.

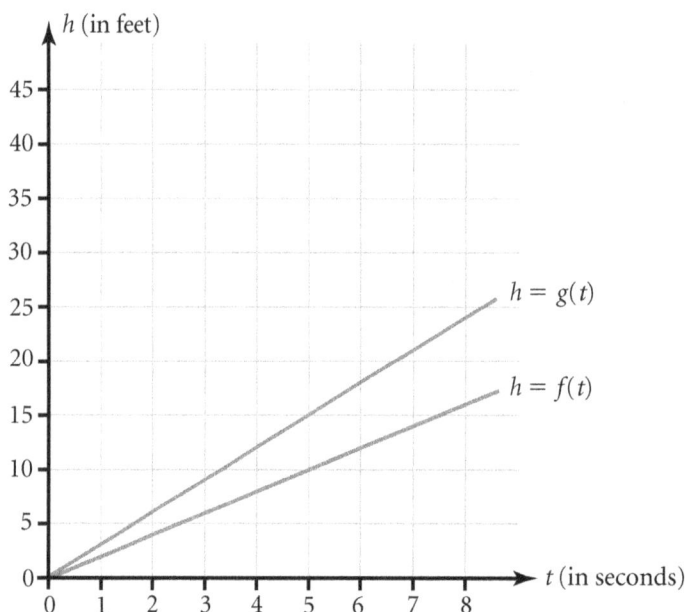

   a. What does the graph of the function f tell you about the speed at which step S rises?

   b. What does the graph of the function g tell you about the speed at which Gabriel climbs the steps?

   c. Copy the graph. On the same set of axes, draw the graph of the function $h = k(t)$, which shows Gabriel's height after t seconds as he is walking up the moving escalator.

   d. How are the three graphs related? How are the three speeds related?

continued ▶

2. Gabriel is now on a train. As the train starts to accelerate from a stopped position, Gabriel begins to walk toward the front of the train. The graph of the function $d = f(t)$ shows the distance the train has traveled after t seconds. The graph of the function $d = g(t)$ shows how far Gabriel would be from his starting point after t seconds if the train had not been moving.

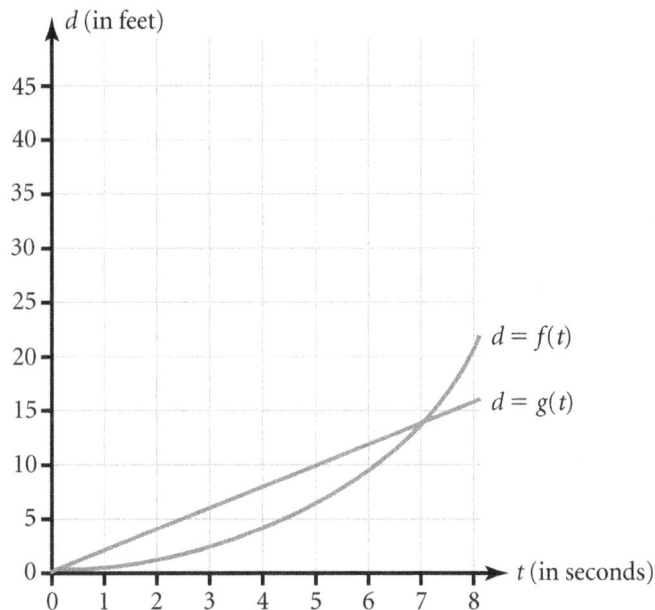

a. What is Gabriel's walking speed?

b. Estimate the train's instantaneous speed at each of these times. In other words, estimate $f'(4)$ and $f'(6)$.

   i. $t = 4$

  ii. $t = 6$

c. Copy the graph. On the same set of axes, draw the graph of the function $d = k(t)$ that shows how far Gabriel is from his starting point after t seconds when he is walking forward through the moving train.

d. Estimate $k'(4)$ and $k'(6)$ from your graph.

e. Discuss how your answers to Questions 2a, 2b, and 2d are related.

Down the Drain

A bathtub is filled with 224 liters of water. When the drain is opened, it takes 4 minutes 40 seconds to empty the tub completely.

1. Assuming the water drains at a constant rate, what is that rate?

2. a. Write a function showing the amount of water $W(t)$ in the tub at t seconds after the drain is opened.

   b. Create a graph of your function.

3. a. Find the derivative function $W'(t)$.

   b. Create a graph of the derivative function.

4. Explain how you could use the graph of the accumulation function to construct the rate function.

5. a. If you started with the graph of the rate function, what else would you need to know, in order to construct the accumulation function?

   b. If you were given this information, explain how you could use the graph of the rate function to construct the accumulation function.

Zero to Sixty

Your new car can go from a stopped position to 60 miles per hour in 10 seconds. Now you will analyze how far your car travels as a function of time during that 10-second interval.

Because you're working with time measured in seconds, use the fact that a speed of 60 miles per hour is equal to 88 feet per second.

1. Explain why 60 miles per hour is the same speed as 88 feet per second.

Assume your car's speed is increasing at a constant rate. After t seconds, you will be traveling $8.8t$ feet per second.

2. Make a graph showing your speed as a function of time.

3. Use your speed graph to find the total distance you travel during each time interval.
   a. From $t = 0$ to $t = 2$
   b. From $t = 2$ to $t = 4$
   c. From $t = 4$ to $t = 6$
   d. From $t = 6$ to $t = 8$
   e. From $t = 8$ to $t = 10$

Polynomial Derivatives

For many functions, you can find the derivative by applying some general principles and using the known derivatives of a few basic functions. This is especially true for polynomial functions, as in this activity. The process of finding a derivative is often called **differentiating** the function.

1. Differentiate each function. That is, find y' as an expression in terms of x.

    a. $y = 5x^3$

    b. $y = 6x^2$

    c. $y = 3x$

    d. $y = 3x^3 - 5x^2 + 8$

    e. $y = x^4 + 3x - 4$

    f. $y = x^4 + 3x + 7$

Once you're comfortable with the steps of finding the derivative of a polynomial function, you are ready to do the reverse process: finding the **antiderivative.**

2. Use Questions 1e and 1f to explain why functions have more than one antiderivative.

3. Find one antiderivative for each of these functions. Use $F(x)$ to represent the antiderivative.

    a. $y = x^3$

    b. $y = 2x^2$

    c. $y = 5x$

    d. $y = x^2 + 2x - 5$

    e. $y = 2x^3 - x + 1$

4. Choose one example from Question 3 and find two more antiderivatives for that function.

Area and Distance

You've seen that if the graph of a function shows the speed at which an object is traveling, then the area under the graph from time 0 to time t will give the accumulated distance traveled during that interval. You will now apply this idea to the situation of a falling object, in which speed is changing at a constant rate.

This graph shows speed as a function of time, based on the equation $s = 32t$. In other words, it describes the speed of a falling object with constant acceleration, gaining 32 feet per second in speed every second.

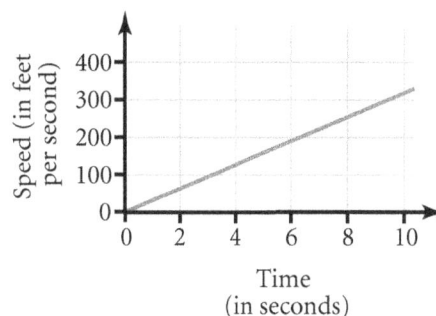

1. Explain the relationship between speed and acceleration in this situation.

2. Explain, based on the idea of antiderivatives, why the function for the distance fallen in t seconds is $d(t) = 16t^2$.

3. Without referring to your work from Question 2, find the area under the graph for each of the given intervals.
   a. $t = 0$ through $t = 0$
   b. $t = 0$ through $t = 2$
   c. $t = 0$ through $t = 5$
   d. $t = 0$ through $t = 8.5$
   e. $t = 0$ through $t = 10$

4. Explain why your answers to Question 3 are consistent with your result from Question 2.

5. Using your data from Question 3, create a graph of the distance $d(t)$ that the object falls in t seconds. Describe the characteristics of this graph.

A Fundamental Relationship

This graph, from the activity *Area and Distance*, shows the speed of a falling object as a function of time, based on the equation $s = 32t$.

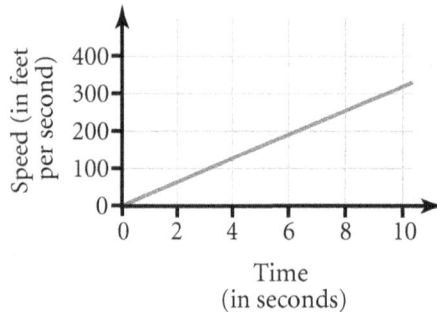

1. Consider the portion of the graph near where $t = 2$, that is, about 2 seconds after the object has been dropped from rest.

   a. Use the area under the graph to calculate how far the object falls between 1.5 seconds and 2 seconds after release. Then use this distance, along with the elapsed time, to calculate the object's average speed over this interval.

   b. Repeat part a for the interval between 1.9 seconds and 2 seconds.

   c. Repeat part a for the interval between 2 seconds and 2.1 seconds.

   d. Based on only these results, what would you estimate the object's instantaneous speed to be at exactly 2 seconds?

   e. How can you verify the accuracy of your estimate?

2. Now consider the portion of the graph near where $t = 10$, that is, about 10 seconds after the object has been dropped from rest.

   a. Use the area under the graph to calculate how far the object falls between 9.9 seconds and 10 seconds after release. Then use this distance, along with the elapsed time, to calculate the object's average speed over this interval.

   b. Repeat part a for the interval between 9.99 seconds and 10 seconds.

continued ▶

c. Repeat part a for the interval between 10 seconds and 10.01 seconds.

d. Based on your results, what do you estimate the instantaneous speed to be at exactly 10 seconds?

e. How can you verify the accuracy of your estimate?

3. Based on your findings in Questions 1 and 2, clearly describe two ways you could find the object's instantaneous speed at any given time.

4. Let $d(t)$ represent the function measuring accumulated distance from time 0 to any later time t on the speed graph. Let $s(t)$ represent the function measuring instantaneous speed at time t. Describe the relationship between $d(t)$ and $s(t)$.

5. Now consider the abstract function $y = f(x)$ shown here. Define the function $A(x)$ to be the shaded area under the graph from $x = 0$ to any positive value of x. Note that this area grows as x increases. Recall the estimates of area (distance) and rate (speed) that you made in Questions 1 and 2.

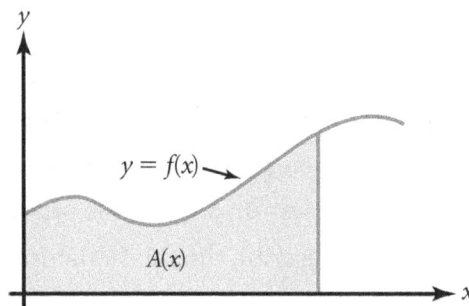

   a. Suppose x increases by a small amount h, as shown in the graph. Estimate the corresponding amount of change in the area under the graph.

   b. Estimate the rate at which this area is growing (with respect to x) at this particular value of x.

   c. As h gets smaller and smaller, your estimate will get closer and closer to the instantaneous rate of change of the area. By definition, this is the derivative $A'(x)$ at that value of x. What is the relationship between the functions $f(x)$ and $A(x)$?

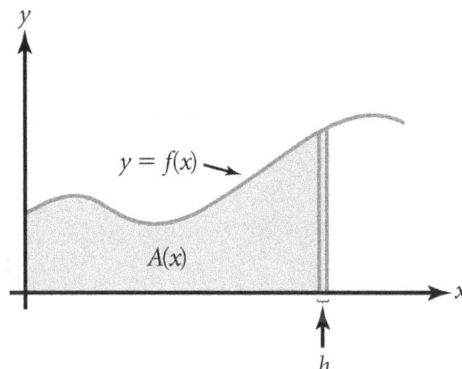

continued ⬤

6. Here is the same arbitrary function $f(x)$ from Question 5. This time, however, the function $A(x)$ describes the shaded area under the graph starting from an arbitrary constant value $x = k$ (not necessarily 0) up to some value of the variable x. Note that k is fixed but $A(x)$ varies as x varies.

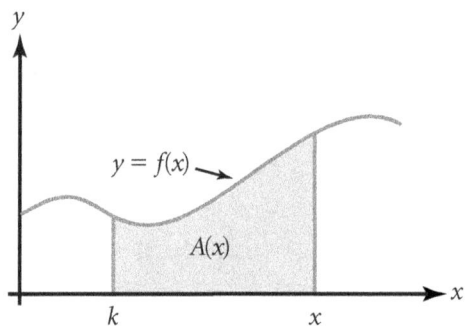

a. Does the area function $A(x)$ have the same values as the function $A(x)$ in Question 5? Explain.

b. Does the derivative function $A'(x)$ have the same values in Question 5 as in this question? Explain.

c. Describe the relationship between $A(x)$ and $f(x)$ symbolically.

d. Describe the relationship between $A(x)$ and $f(x)$ in words.

7. In Question 6, you described a very general and fundamental relationship between a function and the area under its graph from a fixed starting point. State this relationship using the language of rate and accumulation

The Leading Edge

You will soon return to the first unit problem, which involves the volume of a pyramid. In preparation, explore these problems involving simpler geometric forms, their sizes, and how they "grow." Be sure to attach proper units to all quantities.

1. A tractor is pulling a mower through a wheat field. The tractor is traveling in a straight line. It has already mowed a 6-foot-by-9-foot rectangular patch.

   9 ft.

   6 ft.

   a. Suppose the mower is 6 feet wide. If the mower moves forward 1 foot, by how much does the area of the mowed patch increase? At what rate is the area of this region growing? Measure the rate with respect to the change in length, that is, "per foot."

   b. Now suppose the mower is 9 feet wide. If the mower moves forward 1 foot, by how much does the mowed patch increase? At what rate is this rectangular region growing? Again, measure the rate in units per foot of growth.

   c. Try to generalize your results to any rectangle. If the length of one side of the rectangle is increasing, at what rate is the area growing?

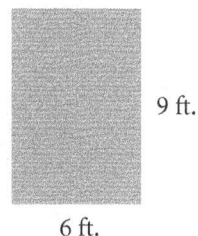

2. Eric built a small storage shed in his backyard. The shed is 8 feet wide, 5 feet deep, and 7 feet high, with a flat top. Now Eric wants more storage space, so he plans to expand the shed. He is trying to decide whether to extend the front forward, the side sideways, or the top upward.

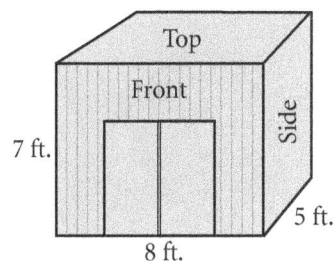

   Top

   Front

   Side

   7 ft.

   5 ft.

   8 ft.

   a. For each of these three possible ways to expand, determine the rate at which the volume of the shed would increase. As in the mowing problem, measure this rate with respect to the change in length of the given dimension—width, depth, or height.

   b. Once Eric has decided on which direction to expand, describe how he can calculate the resulting increase in the shed's volume.

continued ▶

3. In the most common type of automobile engine, pistons move up and down in cylinders as part of the internal combustion cycle. As a piston moves downward, it leaves a space. (This space will be filled with a mixture of gasoline and air, but ignore this for now). That is, as the piston is moving downward, the empty portion of the cylinder is increasing in size.

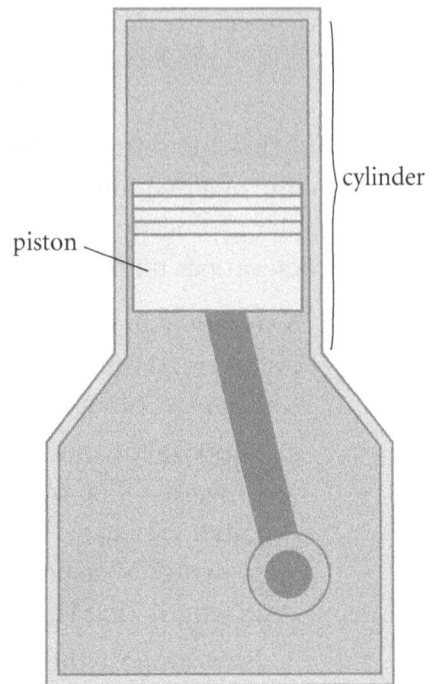

a. Suppose a given cylinder has a diameter of 8 centimeters. As the piston moves downward, at what rate is the volume of the empty space increasing with respect to the distance traveled by the piston?

b. For a cylinder of radius r, at what rate is the volume of the empty space increasing with respect to the distance traveled by the piston?

c. If the piston in part b moves downward x centimeters, by what amount does the volume of the empty space change?

4. Questions 1 to 3 all deal with the rate of growth of an area or volume of a geometric figure. Describe any other common features you see in all three situations.

Pyramids and Energy

You now have most of the tools you need to solve the two main problems of this unit: finding the volume of a pyramid and determining how a solar collector accumulates energy. On your way to completing your solutions to these problems, you will learn more about angle measures and derivatives.

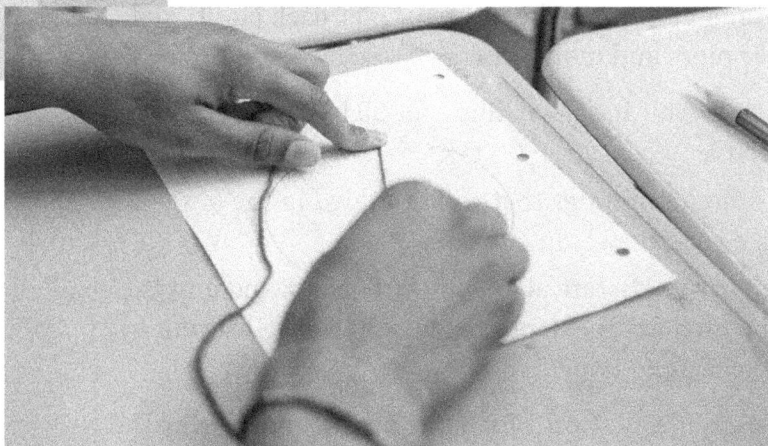

To help solve the solar energy problem, you will explore new ways to measure circles and angles.

Filling the Reservoir

A large water reservoir is in the shape of an inverted pyramid, so that its point is on the bottom. The reservoir is the same shape as the pyramid in the opening activity, *Building the Pyramid*. The base of the reservoir (which is at the top) is a square 100 feet on each side. The reservoir is 100 feet deep.

The reservoir is being filled from an outside source in such a way that the depth of the water is growing at a constant rate of 1 foot per hour. Adding a foot of depth doesn't take much water at first. As the reservoir fills, each additional foot of depth requires more and more water.

1. Imagine that the reservoir has been filling for 30 hours, so the water is now 30 feet deep.
   a. Estimate the amount of water that will be added in the next 5 minutes. Give your answer in cubic feet.
   b. What is the approximate rate at which water is flowing into the reservoir at the moment the depth reaches 30 feet? Give your answer in cubic feet per hour.

2. Go through the steps in Question 1 for some other times during the filling process. You might use time intervals other than 5 minutes.

3. Look for a general rule for the rate at which water is flowing into the reservoir as a function of the time elapsed since the filling process began.

A Pyramid of Bright Ideas

You have just solved the first unit problem, finding the volume of a particular pyramid.

1. Review the concepts of rate and accumulation and the connection between them.

   Now summarize the key ideas of the reasoning that led to the solution of the pyramid problem. In particular, discuss how you used the idea of "filling the reservoir" to find the volume of water in the full reservoir.

You will now turn your attention to the second unit problem, finding the total energy collected by a solar panel in a day.

2. First, reread the problem in *Warming Up*. Second, review the ideas you already developed about this situation, including your graphs from *Warming Up* and from the class discussion of *Total Heat*. Third, think about what you've learned so far in this unit.

   Now write out a strategy for how to complete the solution of the *Warming Up* problem. Be as specific as you can. If any of the steps in your strategy require something you don't yet know how to do, try to describe what it is you need to learn.

Trying a New Angle

You will now explore some new ways of measuring circles and angles. The results will be very useful in your solution of the solar energy problem.

Exploration

1. Make a copy of the unit circle shown here. Suppose you had a piece of string the same length as the radius (that is, 1 unit long).

   a. To the nearest whole number, approximately how many of these unit lengths would you need to wrap all the way around the circle?

   b. Approximately what fraction of the circumference would one of these radius units cover? Darken this portion of your circle (called an **arc**).

   c. Draw a second radius so that the **central angle** formed by the two radii *intercepts* (exactly includes) the darkened arc. Approximately how many degrees is this angle measure?

2. Sketch a circle of arbitrary radius r. Repeat all three parts of Question 1 for this general circle.

3. Repeat Question 1 again, using a unit circle. This time, however, give exact values rather than approximations.

4. If you repeated Question 3 using an arbitrary circle of radius r, would you get the same or different answers? Why?

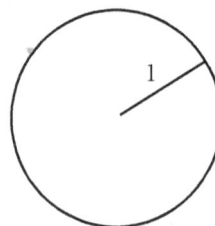

Consolidation

The regularities you observed lead to a new system of angle measurement. The measure of the central angle you just created is the unit of this system. It is called a **radian.**

5. Try to write a clear definition of *radian,* using appropriate vocabulary.

continued ▶

6. Suppose two radii are drawn in a circle of radius r, forming a central angle θ.

   a. If θ is given in degrees, how could you convert its measure to radians? Write an equation to do this.

   b. If θ is given in radians, how could you convert its measure to degrees? Write an equation to do this.

7. Radian measure can also be used to calculate the length of an arc.

   a. A central angle of $\frac{\pi}{8}$ radians is drawn in a circle of radius 25 centimeters. What is the length of the arc made by this angle?

   b. Suppose a circle of radius r contains a central angle of θ measured in radians. How could you find the length of the arc made by the angle? Write an equation.

Application

Now you will practice applying the relationships you've just developed. Be sure your calculator is in the proper mode (degrees or radians) for a given calculation.

8. What is the radian measurement for an angle of 45°? An angle of 120°? An angle of 328°? Give each answer twice: once in terms of π and once as a decimal approximation.

9. What is the degree measurement for an angle of $\frac{\pi}{6}$ radians? Of $\frac{5\pi}{4}$ radians? Of $\frac{10\pi}{3}$ radians? Of 4 radians?

10. A blueberry pie is baked in a 12-inch-diameter pie pan. If you cut an ordinary slice of pie with an angle of 60° at the tip, what would be the length of the edge of the crust?

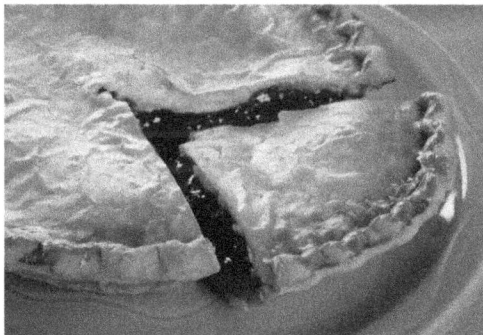

Different Angles

This activity reinforces what you've learned about radian measure and extends it to trigonometric functions, paving the way toward a solution for the *Warming Up* unit problem. Whenever you use your calculator, be sure it's in the proper mode (degrees or radians).

1. What is the radian measure of a straight angle (which marks an arc that is half a circle)? Of a complete revolution (a full circle)?

2. What is the radian measure for an angle of 60°? An angle of 30°? An angle of 300°?

3. What is the degree measure for an angle of $\frac{3\pi}{4}$ radians? Of $\frac{\pi}{12}$ radians? Of $\frac{7\pi}{6}$ radians?

4. In a circle of radius 50, what is the length of the arc made by a central angle of $\frac{4\pi}{3}$ radians? Of 26 degrees?

5. Evaluate each expression.
   a. $\sin 30°$
   b. $\sin\left(\frac{\pi}{6}\right)$
   c. $\cos\left(\frac{3\pi}{4}\right)$
   d. $\sin 5$

6. Sketch one full cycle of the graph of the function $y = \sin x$. Use radian measure for x, so your graph goes from $x = 0$ to $x = 2\pi$. Use the same scale for the vertical and horizontal axes.

7. Draw the line that is tangent to your graph at the origin.
   a. Estimate the angle this line makes with the horizontal axis.
   b. Estimate the slope of this line.

8. Describe how your answers to Questions 6 and 7 would be different if you used degrees instead of radians.

A Solar Formula

The activity *Warming Up* describes the behavior of a solar-collection system. The graph here shows the rate at which that system is gaining energy over the course of a day. The power is 0 between midnight and 6 a.m. and again between 6 p.m. and midnight, because the system does not gain energy during those time periods.

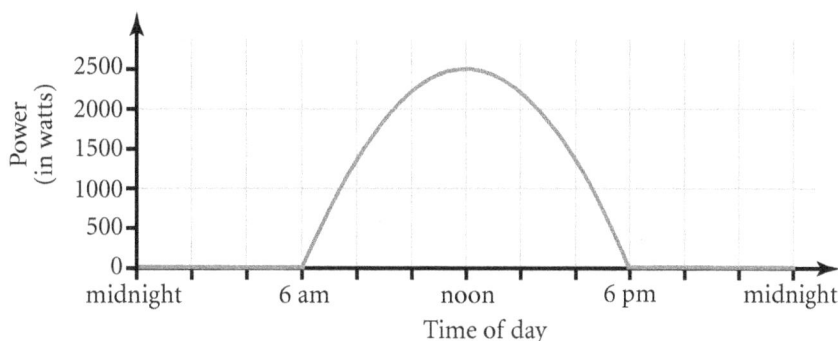

Your task is to develop a specific equation for the power, *P*, in terms of time, *t*, for the period from 6 a.m. to 6 p.m.

Here are two things you can do to simplify your task:

- Use $t = 0$ to represent 6 a.m., so that the graph covers the range from $t = 0$ to $t = 12$.

- Remember that using degrees introduces an arbitrary and confusing unit into the graph, whereas radians are pure numbers and easier to work with. Consequently, use radians to calculate values for the sine function.

A Sine Derivative

To get an equation for the accumulated energy from the solar power system, you need to find an antiderivative for the function that describes the rate at which the system gains energy. Because that rate involves the sine function, and because the accumulation graph itself looks something like the sine function, a good place to start looking for the antiderivative is to examine the derivative of the sine function.

The graph shows one full cycle of the function $y = \sin x$, using radian measure and using the same scale for the vertical and horizontal axes. The horizontal scale shows key values in terms of π.

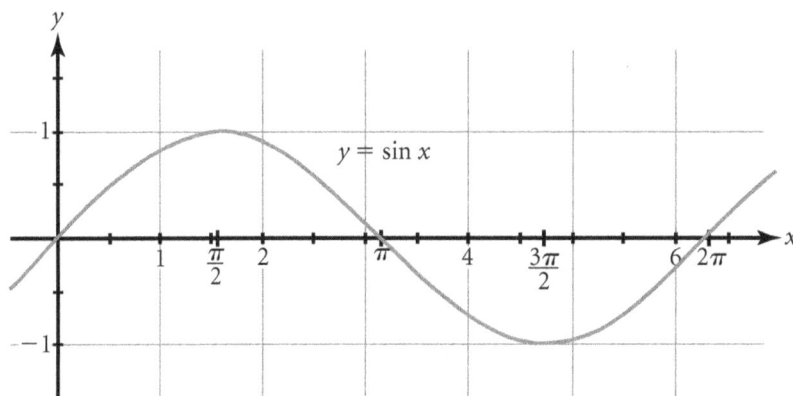

1. Using the graph, estimate the derivative for each x-value.

   a. $x = 0$ b. $x = \dfrac{\pi}{4}$ c. $x = \dfrac{\pi}{2}$

   d. $x = \dfrac{3\pi}{4}$ e. $x = \pi$ f. $x = \dfrac{5\pi}{4}$

   g. $x = \dfrac{3\pi}{2}$ h. $x = \dfrac{7\pi}{4}$ i. $x = 2\pi$

2. Plot your results from Question 1 using the same scale as the graph for $y = \sin x$.

3. What function does your graph suggest for the derivative of the sine function?

A Derivative Proof

Finding the derivative of the function $y = \sin x$ at a point $(x, \sin x)$ involves looking at the ratio $\sin \dfrac{(x + h) - \sin x}{h}$. Your task is to examine what happens to this derivative ratio as h gets smaller and smaller. The method uses the ideas that x represents an angle and that the trigonometric functions are defined in terms of right triangles.

The diagram shows a portion of a unit circle, with a central angle labeled x and another central angle labeled h. You will be looking at what the derivative ratio $\sin \dfrac{(x + h) - \sin x}{h}$ means in terms of this diagram and what happens to this ratio as h gets smaller. The first part of your work will be to justify some of the labels in the diagram.

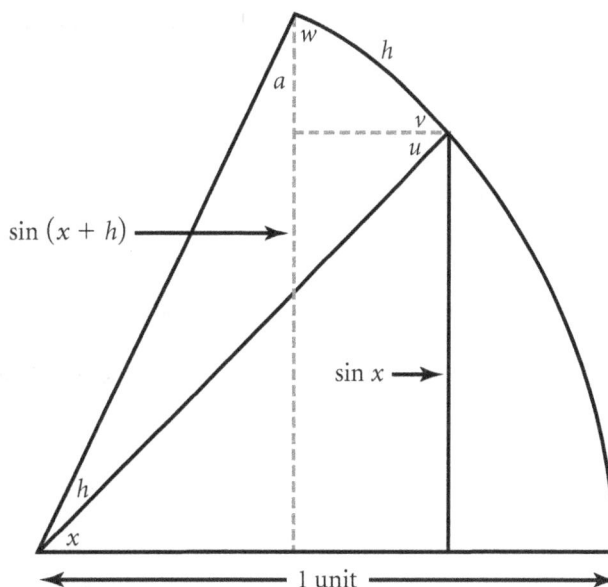

1. The solid vertical line in the diagram is labeled $\sin x$. Why is this label correct?

2. The long dashed line in the diagram is labeled $\sin (x + h)$. Why is this label correct?

3. The diagram uses the letter h for the small arc of the circle as well as for the small central angle. Why is it correct to use the same variable for these two parts of the diagram?

If the angle h were very small, the curve of the small arc h would be negligible. In the rest of this activity, think of this arc as a line segment, so that the upper portion of the diagram shows a very small right triangle with hypotenuse h.

continued ▶

4. Use the labeling discussed in Questions 1 and 2 to get an expression for the vertical side a of this small triangle.

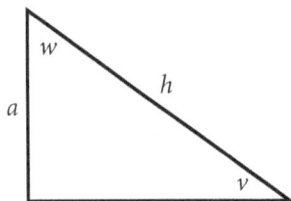

5. Use your answer to Question 4 to rewrite the ratio $\dfrac{\sin (x + h) - \sin x}{h}$ as a simpler expression.

6. Consider the angles labeled x, u, v, and w. How are they related? (*Suggestion:* When you think about the relationship between angles u and v, keep in mind that a line tangent to a circle at a point is perpendicular to the radius to that point.)

7. Combine your answer to Question 5 with your results from Question 6 to express the ratio $\dfrac{\sin (x + h) - \sin x}{h}$ in terms of only x.

8. Explain how your work in this activity demonstrates that the derivative of the sine function is the cosine function.

A Cosine Derivative

So, the derivative of the sine function is the cosine function! And what's the derivative of the cosine function? Here's your chance to make a conjecture.

The graph shows one full cycle of the function $y = \cos x$, using radian measure and the same scale for the vertical and horizontal axes. The horizontal scale shows key values in terms of π.

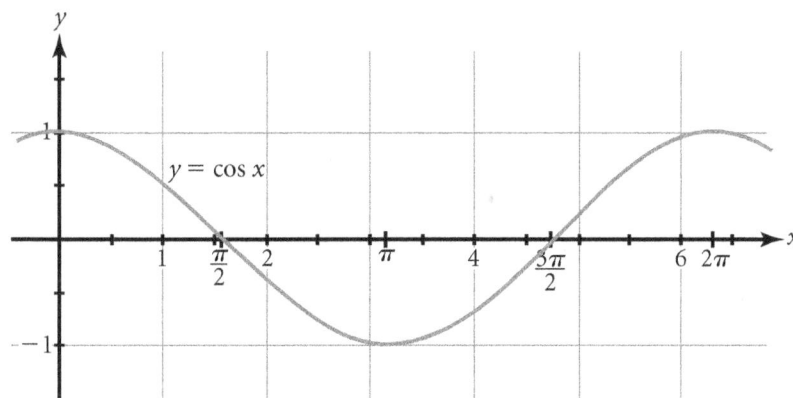

1. Using the graph, estimate the derivative for each x-value.

   a. $x = 0$

   b. $x = \dfrac{\pi}{4}$

   c. $x = \dfrac{\pi}{2}$

   d. $x = \dfrac{3\pi}{4}$

   e. $x = \pi$

   f. $x = \dfrac{5\pi}{4}$

   g. $x = \dfrac{3\pi}{2}$

   h. $x = \dfrac{7\pi}{4}$

   i. $x = 2\pi$

2. Plot your results from Question 1, using the same scale as the graph for $y = \cos x$.

3. What function does your graph suggest for the derivative of the cosine function?

The Inside Story

If you start with the function $y = -\cos x$, you get the derivative $y' = \sin x$. But for the solar energy problem, you need a function whose derivative is $2400 \sin\left(\frac{\pi}{12}x\right)$, which has an "inside coefficient" of $\frac{\pi}{12}$.

It's a reasonable guess that the desired antiderivative might involve $\cos\left(\frac{\pi}{12}x\right)$. To explore this idea, you will examine the simpler situation of the function $y = \cos(2x)$.

The graph shows one full cycle of the function $y = \cos(2x)$. Because of the inside coefficient of 2, a full cycle starting at $x = 0$ ends at $x = \pi$.

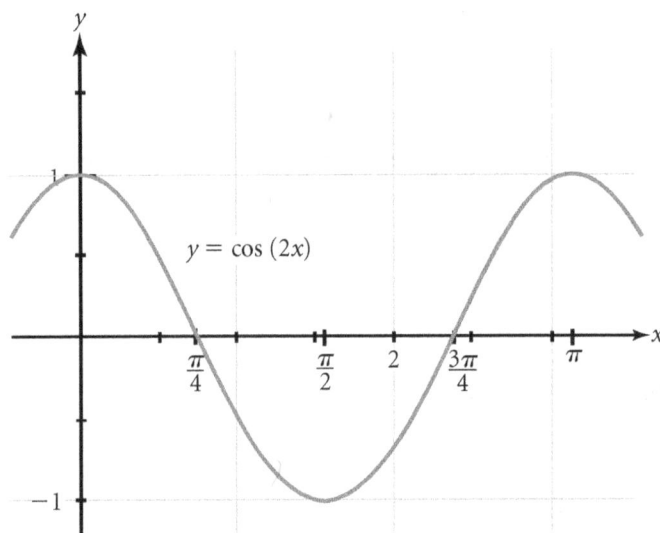

$y = \cos(2x)$

1. Using the graph, estimate the derivative for each x-value.

   a. $x = 0$ b. $x = \frac{\pi}{8}$ c. $x = \frac{\pi}{4}$

   d. $x = \frac{3\pi}{8}$ e. $x = \frac{\pi}{2}$ f. $x = \frac{5\pi}{8}$

   g. $x = \frac{3\pi}{4}$ h. $x = \frac{7\pi}{8}$ i. $x = \pi$

2. Plot your results from Question 1, using the same scales as the graph for $y = \cos(2x)$.

3. What function does your graph suggest for the derivative of the function $y = \cos(2x)$?

A Solar Summary

You began examining the solar-collector problem in *Warming Up*, and you estimated the total energy for a single day in *Total Heat*. You have now applied ideas about derivatives and antiderivatives to get a precise value for that total energy.

Explain the steps involved in getting your result. Make sure to discuss the role of derivatives and antiderivatives in this process, as well as the insights you used about trigonometric functions.

How Much? How Fast? Portfolio

You will now put together your portfolio for *How Much? How Fast?*
This process has three steps:

• Writing a cover letter summarizing the unit

• Choosing papers to include from your work in this unit

• Discussing your personal mathematical growth in the unit

Cover Letter

Look back over *How Much? How Fast?* and describe the two central
problems of the unit and the key mathematical ideas. Your description
should give an overview of how the key ideas were developed and
how they were used to solve the central problems. These key ideas
include extending the sine and cosine functions and writing functions
describing various aspects of the motion of a falling object.

In compiling your portfolio, you will select some activities you think
were important in developing the unit's key ideas. Your cover letter
should include an explanation of why you selected each item.

Selecting Papers

Your portfolio for *How Much? How Fast?* should contain these items:

• One or two activities that helped you understand the relationship
  between rate graphs and accumulation graphs

• *A Pyramid of Bright Ideas*

• *A Solar Summary*

Personal Growth

Your cover letter for *How Much? How Fast?* describes how the
mathematical ideas develop in the unit. In addition, write about your
own personal development during this unit. You may want to address
this question:

> *How does your work in this unit compare with any previous
> impressions you had about the subject of calculus?*

Include any other thoughts about your experiences that you wish to
share with a reader of your portfolio.

Supplemental Activities

The supplemental activities for this unit extend the ideas you've been learning about derivatives, area, and functions. Here are two examples:

- *Parabolic Area* follows up your study of the fundamental theorem of calculus with a specific example.

- *Ana on the Train* blends several ideas from the unit—including accumulation, estimation, derivatives, and the fundamental theorem of calculus—into a single problem.

Derivative Power

You've seen that these three statements are true:

- If $y = x^2$, then $y' = 2x$.
- If $y = x^3$, then $y' = 3x^2$.
- If $y = x^4$, then $y' = 4x^3$.

1. Using these examples as a model, make a conjecture about the value of y' if $y = x^n$.

2. Prove your conjecture using algebra. Assume n is a positive integer. You might want to use the binomial theorem, which explains the coefficients in the expansion of an expression of the form $(a + b)^n$.

3. *Challenge:* Try to prove your conjecture for the cases of $x = -1$ and $x = -2$.

Parabolic Area

Take a look at the diagram. Your main task in this activity is to find the value of the shaded area under the graph of the equation $f(x) = x^2$. The diagram also shows two general vertical lines at $x = t$ and $x = t + h$.

Define a function $A(t)$ as the area under the graph of $f(x)$ from $x = 0$ to the line $x = t$. The definition of derivative says that $A'(t)$ is the ratio $\frac{A(t + h) - A(t)}{h}$.

1. Use the diagram and the definition of the function A to express the numerator $A(t + h) - A(t)$ as an area.

2. Based on your answer, explain why the derivative ratio $\frac{A(t + h) - A(t)}{h}$ is approximately equal to $f(t)$.

3. Explain why your approximation in Question 2 gets even better as h gets smaller.

4. Use your results from Questions 1 to 3 to explain why $A'(t) = f(t)$, so A is an antiderivative of f.

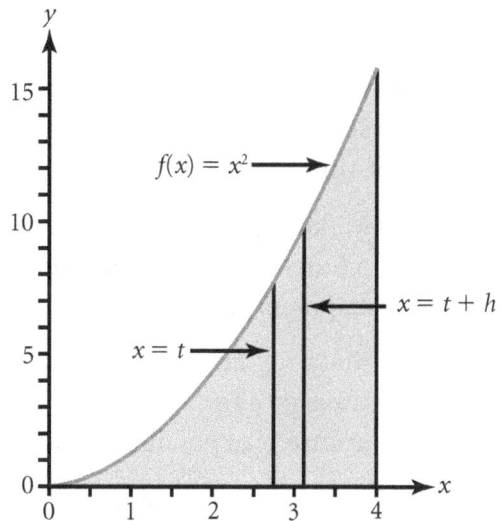

5. Use your conclusion from Question 4 to find an algebraic expression for $A(x)$.

6. Explain why the shaded area is equal to $A(4)$. Then use your answer to Question 5 to find the numeric value for this area.

Ana on the Train

Ana has an accelerometer that measures her acceleration in feet per second per second. While riding on a train, Ana measured her acceleration over 15-second time intervals. Although the accelerometer fluctuated a bit over each time interval, Ana estimated the average value for each interval.

Time (minutes:seconds)	Acceleration (feet/s/s)
0:15	3.2
0:30	0.9
0:45	2.0
1:00	0.0
1:15	0.0
1:30	−0.5
1:45	0.0
2:00	−1.2
2:15	−1.2
2:30	−3.2

As you work, treat each value in the table as if the acceleration were constant for the period ending at that time.

1. Make a graph of Ana's velocity as a function of time. Be as accurate as possible.

2. Make a graph of Ana's distance traveled as a function of time. Be as accurate as possible.

3. Your distance graph is an accumulation graph. Find the instantaneous slope of this graph at two times: 1:10 and 2:10. Do this by drawing tangent lines and calculating their slopes from the graph. Label these slopes with the correct units.

4. State the **fundamental theorem of calculus** and explain its significance in connection with Question 3.

5. Find an equation for the distance function for each time interval.

   a. $1:15 < t < 1:30$

   b. $2:00 < t < 2:15$

Widget Wisdom

Evan makes widgets, but his production rate is uneven. The graph shows how many widgets Evan makes, over the course of an 8-hour workday, as a function of the time elapsed. For instance, $f(6) = 375$, which means that at the end of 6 hours, he has made 375 widgets.

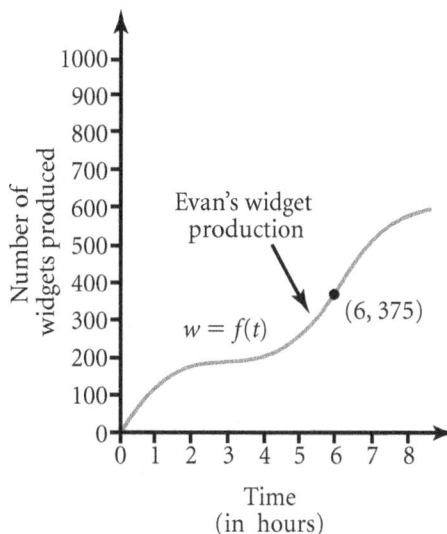

1. Based on this graph, estimate the rate at which Evan is making widgets at $t = 6$.

Evan has two co-workers, Cassandra and Liam, who are both more productive than Evan, though in different ways.

For every widget Evan produces, Cassandra produces two widgets. For example, by the end of 6 hours, Cassandra has produced 750 widgets, instead of only 375.

2. Make a copy of the graph. On the same set of axes, sketch a graph showing Cassandra's widget production.

3. Estimate the rate at which Cassandra is making widgets at $t = 6$.

Liam accomplishes what Evan does in one-third the time. For example, it takes him only 2 hours to produce the 375 widgets for which Evan needs 6 hours. Because of his speed, Liam works for only 2 hours 40 minutes.

4. On your axes, sketch a graph showing Liam's widget production.

5. Estimate the rate at which Liam is making widgets at the moment he creates his 375th widget.

Photographic Credits

Front Cover Photography

(upper row) Stephen Loewinsohn; (lower row and background image) iStockphoto

How Much? How Fast?

1 (upper row) Stephen Loewinsohn; (lower row and background image) iStockphoto; **3** Stephen Loewinsohn; **4** Shutterstock; **5** Shutterstock; **11** iStockphoto; **12** Shutterstock; **13** Shutterstock; **14** iStockphoto; **17** iStockphoto; **18** Stephen Loewinsohn; **21** Shutterstock; **28** Shutterstock; **29** iStockphoto; **36** Shutterstock; **37** Stephen Loewinsohn; **38** iStockphoto; **39** Shutterstock; **41** iStockphoto; **46** Shutterstock; **54** iStockphoto

www.ingramcontent.com/pod-product-compliance
Lightning Source LLC
Chambersburg PA
CBHW062029210326
41519CB00060B/7361